Dick and Margaret Stokes
408 Greenfield Drive
Mt Pleasant, MI 48858

S0-EIF-375

The Animal Kingdom

ANIMAL DEFENSES

Malcolm Penny

Illustrated by Carolyn Scrace

The Bookwright Press
New York · 1988

The Animal Kingdom

Animal Camouflage
Animal Defenses
Animal Evolution
Animal Homes
Animal Migration
Animal Movement
Animal Reproduction
Animals and their Young
Endangered Animals
The Food Chain
Hunting and Stalking

First published in the
United States in 1988 by
The Bookwright Press
387 Park Avenue South
New York, NY 10016

First published in 1988 by
Wayland (Publishers) Ltd
61 Western Road, Hove
East Sussex BN3 1JD, England

© Copyright 1988 Wayland (Publishers) Limited

Library of Congress Cataloging-in-Publication Data

Penny, Malcolm.
  Animal defenses/by Malcolm Penny: [illustrated by Carolyn Scrace].
  p. cm. — (The animal kingdom)
  Bibliography: p.
  Includes index.
  Summary: Discusses methods of animal defense, including shells, armor, poison, bites, camouflage, and playing dead. Includes a section on animals which need help in being defended.
  ISBN 0-531-18192-8
  1. Animal defenses—Juvenile literature. [1. Animal defenses.]
I. Scrace, Carolyn, i11.   II. Title.   III. Series: Penny, Malcolm.
Animal kingdom.
QL 759.P46 1988
591.57— dc 19

87-32453
CIP
AC

Typeset by DP Press, Sevenoaks, Kent, England
Printed by Casterman SA, Belgium

**Words printed in bold in the text are
explained in the glossary on page 30.**

Contents

Sensing danger	4
Getting away	6
Hiding from the enemy	8
Spines and prickles	10
Shells and armor	12
Frightening the enemy	14
Bites, stings and shocks	16
Emergency tactics	18
Playing dead and other tricks	20
Safety in numbers	22
Fighting back	24
Defense within the species	26
Animals we need to defend	28
Glossary	30
Further information	31
Index	32

Sensing danger

Many animals spend their lives in danger of being caught and eaten by **predators**. Some of them, for example the very small animals that live in **plankton**, have no defense against their enemies. However, there are usually so many of them that some will survive to breed and carry on the species, even though most of them are eaten.

Most larger animals have some means of defending themselves—perhaps by hiding, or by running away, or sometimes by fighting back when they are attacked.

The first line of defense for a **prey** animal is to be alert to the approach of danger. The most useful senses for this are sight, hearing and smell. A rabbit is a good example of a prey animal that uses all three senses to warn it when a predator is approaching.

Like many other prey animals, rabbits have their eyes at the sides of their heads, not in the front. This gives them a very wide field of vision, so they can see danger approaching not only from the front and the sides but even from behind. They have long ears, which can turn to listen to any suspicious sound, so rabbits can judge very quickly in which direction the danger lies. Rabbits also use their **acute** sense of smell, as they sniff the breeze for the scent of a fox or a weasel, or, indeed, a human being.

Some animals rely on one sense more than another to warn them of danger. Fish usually use their eyes, while small **nocturnal** animals rely more on hearing or scent.

Once an animal has sensed danger, it must react to it, if it is to survive. In this book we shall look at many of the interesting ways in which animals defend themselves against their enemies.

Opposite *The keen senses of such prey animals as rabbits alert them to the approach of foxes, birds of prey and other predators. The rabbit in the distance has seen a buzzard and is attempting to escape from it.*

Getting away

Once prey animals have detected the approach of danger, the best chance of survival for most of them is to escape. Rabbits and mice run for cover, frogs jump into water or long grass, birds and many insects take to the air, while fish swim to a hiding place as fast as they can.

For jumping and flying creatures, the first movement is the most important. Small birds take off very quickly, almost like grasshoppers, jumping off the ground and flying at once. Frogs make one big leap for safety. If they are lucky, their enemy will not see where they land. Otherwise, they must try to escape again but they cannot jump repeatedly without getting tired.

Some antelope of the African plains can surprise the leopard or cheetah that is chasing them by making sudden leaps high in the air. This technique, called "pronking" or "stotting," startles predators, allowing the antelope extra time to escape.

Ground-dwelling tiger beetles can quickly take to the air when danger threatens. In the picture a green tiger beetle has just managed to escape from a female blackbird.

Lobsters and crayfish have an unexpected way of escaping from predators. They use their large, fan-shaped tails to row themselves backward under a rock, with one lightning flick.

Water is used by several creatures as a defense against their enemies. On the ice of the Arctic, slow-moving seals cannot escape polar bears. However, if a seal stays close to a hole in the ice, it can plunge into the safety of the water as soon as it sees a polar bear.

In a much hotter part of the world, the Central American rain forest, lives the basilisk lizard. When chased on land by a predator, the lizard escapes by actually running over the surface of ponds or streams. Its amazing ability to walk on water has earned it the name of the Jesus Cristo lizard.

Some agile South African springbok leap into the air to alarm and confuse the predator that is chasing the herd. This is called stotting or pronking.

Hiding from the enemy

Many kinds of animals do not run away from their enemies but instead stay perfectly still. They can safely do this because their bodies are colored or patterned to look like part of the general background. Such animals are said to be **camouflaged**. This is a good defense because predators do not recognize them as a tasty morsel.

Caterpillars are good examples of camouflaged animals. Most of them are fat, soft and slow-moving, an easy meal for a bird to take to its **nestlings**. To avoid being eaten, many caterpillars are green or brown, so birds mistake them for leaves or twigs. Many moths, stick insects and leaf insects also look more like plants than animals.

Some kinds of frogs, snakes and lizards rely on camouflage to protect them. An example is the leaf-tailed gecko, an Australian lizard. As its name suggests, it looks like a mottled leaf and so fools the birds that would otherwise eat it.

The young of many animals cannot run fast enough to escape a predator, so camouflage is a useful defense for them. Fawns (baby deer) have brown coats dappled with white spots. This looks just like spots of sunlight falling on the ground. Baby birds born in nests on the ground are usually covered with stripes or blotches of gold and brown. Shore birds, gulls and ducklings are all camouflaged in this way.

Birds hatched in nests in hedges and bushes stay under cover whenever they can, hiding among branches. They are often brown colored so they do not show up. Sometimes they venture into the open, where a bird of prey may see them. Then they defend themselves by flying close to the ground, where the hawk is afraid to dive on them.

Opposite *All kinds of animals rely on camouflage to protect them. A young dappled fawn hides among the underbrush, while a young song thrush stays still in an apple tree. Concealed among the leaves of a lime tree is a puss moth caterpillar.*

Spines and prickles

A sure way to put off an enemy and avoid being eaten is to be covered with spines or prickles. Some well-known examples of prickly animals are the hedgehog of Europe, the porcupines of Africa and the United States, and the echidna, or spiny anteater, of Australia and New Guinea.

Each of these prickly animals has its own extra protection. The hedgehog's spines are sharp and, when threatened, the animal rolls up into a spiny ball. The porcupine turns its back on an attacker, so that its long, trailing quills are the first thing to be touched. They are barbed at the end, and quite loosely attached. Any animal that sniffs at a porcupine will come away with sharp quills stuck in its nose. The echidna rolls up like a hedgehog, but it has a secret weapon as well, which we shall discover in a later chapter.

Many small lizards are protected by spines. The little moloch lizard, which lives in central Australia, is sometimes called the thorny devil because it is covered with sharp spines. A very spiny lizard of the United States is the Texas horned lizard, which lives in the Arizona desert.

The animals illustrated here all use spines or prickles to defend themselves against their enemies.

The animals illustrated are not shown to scale

Hedgehog
(Europe)

Moloch lizard
(Australia)

Long-nosed spiny anteater
(New Guinea)

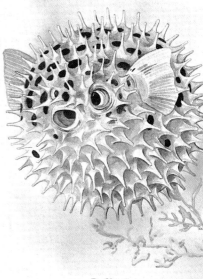

Puffer fish
(Indian Ocean)

Many small fish have a row of spines along the back, which can be used as a means of defense. Sticklebacks are a good example. If one is caught by a pike, it will raise its spines inside the pike's mouth, to keep itself from being swallowed.

Puffer fish on tropical reefs look slightly spiky but otherwise quite normal until they are alarmed. If they are attacked, or if a person comes too close to them, they can suck in enough water to blow themselves up into a ball covered with spines. They suddenly become quite impossible even for a large fish to eat.

Spines need not always be strong and sharp to drive away enemies. Some caterpillars have very fine, hairlike spines. They are hollow and filled with a substance that causes pain or itching inside birds' bills or on people's fingers.

Crested porcupine
(Africa)

Shells and armor

Some prey animals are not covered with spines, but with strong armor, which protects their bodies from predators. There are many armored animals, from snails to **reptiles** to **mammals**.

Some large land snails are well armoured, but for most of them a very thick shell would be too heavy to carry around. The thickest snail shells are found in the ocean. Whelks, cockles, cowries and cones are all protected by thickly armored shells. Only the fish with the strongest jaws, or very large crabs with powerful pincers, can crack them.

The strong shells of tortoises are formed from their ribs, which are joined together and covered with plates of horny skin. Some other reptiles have very strong armor, though their skeletons are normal. Crocodiles and alligators are well armored, and, on a much smaller scale, so are some lizards.

*Illustrated **below** are three different species of armadillos. Besides being protected by their armor, these animals have an extra method of defense. The tiny fairy armadillo and the nine-banded armadillo dive into a hole in the ground if a predator comes too near, while the three-banded is the only species that can roll up into a ball.*

Nine-banded armadillo

Three-banded armadillo

Fairy armadillo

The Australian shingleback lizard and the South African armadillo lizard are both protected by sharp, overlapping scales. If its armor does not deter a predator, the armadillo lizard will roll itself up into a tight ball.

The armadillo lizard is named after the armadillos of North and South America. Like the pangolins of Africa, armadillos are armor-plated mammals that are covered with hard, horny plates that protect them from being bitten. The plates of a pangolin overlap and have very sharp edges. If a pangolin is attacked, it first curls up, and then straightens out suddenly, wounding the predator's nose or paw that was touching it.

From the examples we have looked at, you can see that armored animals are very well defended against their enemies.

An African pangolin feed on grass, well protected by its razor-sharp scales. It would be hard for a predator to attack the pangolin's face or underside, which are protected by the armor.

Frightening the enemy

In this chapter we shall look at some useful methods of defense that animals use to frighten off enemies. By making a sudden movement or sound, some animals are able to startle a predator for a few moments and escape.

A click beetle uses two ways of startling a bird that is about to eat it. As soon as the bird touches it, the beetle folds its legs and falls to the ground, pretending to be dead. A second or two later, it flicks back its head and springs into the air, leaving the surprised bird without its meal.

Many moths have brightly colored hind wings, which are usually hidden by their camouflaged front wings. When a hungry bird or mouse disturbs them, the moths spread their wings, producing a flash of color. This alarms the predator and usually gives the moth just enough time to escape.

An Australian blue-tongued skink displays its bright-blue tongue at a enemy. By hissing at the same time, the skink appears much fiercer than it really is.

A European common toad puffed up in the defensive posture to scare away a predator.

Perhaps the most amazing example of a creature that uses a flash of color to defend itself is the blue-tongued skink of Australia. This lizard can run at top speed, but when cornered by a predator, it will open its mouth to reveal a bright-blue tongue. Sometimes it hisses at the same time.

Some creatures fool their predators by suddenly making themselves look larger and fiercer than they really are. The Australian frilled lizard puffs out a bright frill of skin around its neck and opens its mouth, which is yellow inside. The common toad defends itself against a predator by inflating its body with air. Suddenly the inflated toad seems too big for the predator to manage.

Many animals all over the world use noise to try to scare off an enemy or a creature that alarms them. Harmless animals like kittens and puppies show their teeth and hiss or bark when they are frightened. Chimpanzees also bare their teeth and make a chattering noise. They may look as if they are laughing, but in fact it is a sign of fear.

Bites, stings and shocks

Many small animals bite or squeal in an attempt to frighten their enemies away, but some can actually harm the animal that is attacking them. These are the animals that can produce poison.

Wasps need a poisonous sting to kill prey, in order to feed themselves and their young. They also use their sting as a means of defense. Bees live on nectar and do not need their stings, except to defend themselves. They will kill any insects that invade their hives, such as wasps and certain kinds of moths. Occasionally they will sting humans, but they only do this when threatened.

Toads and many kinds of frogs produce poison from **glands** in their skin, often around their back legs. The poison has a most unpleasant effect on any animal that tastes it. The famous poison arrow frogs of South America can actually cause the death of their attackers because their venom is so strong. Some snakes can also give poisonous bites.

Most poisonous animals are brightly colored, for a very good reason. After one unpleasant experience with them, a predator will learn to recognize their colors and leave them alone.

Skunks, with their bold black and white stripes, give a clear warning of their defenses, even though they are not poisonous. Instead of poison, they produce a smell that all other animals hate.

Even tiny insects like ants and the puss moth caterpillar can defend themselves effectively by producing acid, which causes a painful, but not poisonous, wound to the predator.

The most alarming form of defense is produced by certain fish, such as the electric eel and the torpedo ray. They can deliver powerful electric shocks through the water, a sure way to deter an enemy.

Opposite *A spotted skunk rears up in a handstand before squirting an unpleasant-smelling liquid to warn off a bob-cat. Alarmed by this, the bob-cat hisses defensively. In the foreground, a poisonous rattlesnake shakes its rattle.*

Emergency tactics

Like many Australian lizards, water skinks can shed their tails. In the picture a brolga (an Australian crane) is left with the lizard's thrashing tail, while the water skink manages to escape being eaten. Very little blood is shed as the tail breaks off, and in time a new tail will grow from the stump of the old one.

Some animals have a defense against predators that they use only as a last chance of surviving an attack. They give part of themselves away.

Lizards are well known for losing their tails when they are seized by some other animal. The tail continues to twitch after it has dropped off, so that more often than not the predator pounces on the tail and eats it, while the lizard escapes.

No doubt the lizard finds it harder to move without a tail for a while, but the loss is not as serious as it might seem. The tail is designed to break in a particular place where it is weak. When the tail breaks off, the lizard bleeds very little, and in a few weeks, it will grow a new tail.

Other animals are able to escape a predator, having lost a limb or two, and replace the missing part later. Starfish, spiders and some young insects can grow new limbs that emerge the next time they **molt**. Adult insects cannot do this because they have finished molting.

Moths and butterflies can shed some of their scales or lose part of a wing. The scales on the wings of moths are very loosely attached, so if the moth flies into a spider's web, the scales stick to the web while the moth goes free. Butterflies often have spots that look like eyes on their wings. When a bird attacks a butterfly, it usually goes for the eyespots first, leaving the butterfly with a damaged wing but with its body unharmed. It might fly awkwardly, but at least it can escape alive, perhaps to find a mate and carry on the species.

The tattered, ragged wings of this South American morpho butterfly show that a bird has attacked its eyespots. The function of the eyespots is to distract the bird's attention from the butterfly's head and body, which are much more important to it than the edges of its wings.

Playing dead and other tricks

In the previous chapter we looked at some animals that survive by giving a part of themselves away. Here we shall discover some animals that give themselves up completely to the predator — and survive! They perform an astonishing trick to fool their enemy, pretending to be dead.

The "playing dead" trick is often called "playing possum" because it is used by possums when they are attacked by larger animals. Some harmless snakes, and even foxes or African hyenas, will also suddenly lie down and pretend to be dead. It works because the predators that hunt them will eat only freshly killed meat, not an animal that is already dead.

Darwin's frog of South America plays dead while moving along. When pursued by an enemy, the tiny frog jumps into a stream and floats along motionless like a piece of driftwood.

The opossum of North America defends itself by lying down and playing dead. If the predator is fooled and leaves the opossum, it will soon come to life again!

Another good defense trick is used by a mother lapwing when she sees a predator coming too close to her nest or chicks. Pretending to have broken her wing, she runs away from the nest, dragging the wing on the ground, and making noises as if she really were wounded. The predator follows her, expecting an easy meal. When she has drawn the attacker far enough away, and before it can get close enough to catch her, the mother bird flies away and later returns to her chicks.

A similar trick is played by the enormous ostriches of Africa. To protect the chicks from a dangerous predator, an adult ostrich may pretend to be hurt, or flap its wings and run around calling loudly. Occasionally it may fall to the ground and stir up a sand storm with its legs and wings.

Small, harmless grass snakes are common in many European countries. They have no venom to defend themselves against an enemy, but sometimes curl up and pretend to be dead.

Safety in numbers

A quite different means of defense against predators is used by animals that gather together in large groups, such as herds of oxen, flocks of birds and shoals of fish. With many eyes, ears and noses, a large group is more likely to sense danger. It is also more likely to frighten off a predator.

The first member of the group to see, hear or smell danger can give a warning to the rest of the group. Prairie dogs squeal or whistle, rabbits and kangaroos thump the ground, and elephants trumpet loudly. Most birds give a loud alarm call that is recognized as a warning by other birds, even those of different species.

Some animals use chemicals to warn others in their group. Minnows, which are small fish that are often eaten by bigger fish, release a substance into the water when they are attacked. Other minnows respond to the substance by swimming away. Bees, on the other hand, produce a chemical substance that brings other bees to the place of danger, with their stings ready.

Many small birds draw attention to birds of prey, such as owls or hawks, by **mobbing** them. The small birds fly around the predator, calling loudly and darting in as if to attack it. Often the mob drives the predator away. Willie wagtails in Australia always mob a kookaburra if it comes near their nest. They do this because kookaburras often steal other birds' eggs and chicks. The wagtails have no hope of hurting the kookaburra, which is much larger than they are, but eventually they drive it away.

In the Arctic, musk oxen form themselves into a ring if they are threatened by wolves, standing side by side with their horns outward. The younger members of the herd are kept safely in the middle.

Opposite *A herd of musk oxen in Arctic Canada form a tight, defensive group to warn off two Arctic wolves that are threatening to attack.*

Fighting back

There are animals, such as the bees that we have just looked at, that react to the approach of danger by fighting, instead of fleeing. They use whatever weapons they have, whether they are stings, horns, teeth or hoofs.

The entire horse family, including zebras, have hoofs with which they can injure quite large enemies, but there are some other kicking animals that are just as powerful. Ostriches have a large nail on the middle toe to help them grip the ground when they run. It is also a dangerous weapon against an attacker. The harmless-looking kangaroo can kill people and dogs by kicking. Even giraffes, though they usually run from danger, will kick in self-defense if they cannot escape.

African rhinoceroses defend themselves with their horns, but Asian rhinoceroses still have front teeth, which they use for fighting their enemies.

Antelope, such as oryx, nyala and gemsbok, will turn and fight a lion or a leopard, especially if they have young with them. Their long horns are a useful defense in such cases.

The secret weapon of the Australian echidna is a strong, sharp claw on each hind foot with which it can cut and scratch. The echidna's relative, the duck-billed platypus, also has claws, or spurs, on its hind legs. They are attached to glands that produce poison strong enough to kill a dog.

On the bottom of the world's oceans there lives an animal related to the starfish called the sea cucumber. It feeds on the tiny creatures in the sand or plankton and looks completely defenseless, like a large soft slug. If it is attacked by a crab, however, it can produce long sticky threads, which entangle the crab until it is quite unable to move, while the sea cucumber crawls quietly away.

Zebras have sometimes been known to kick a lion that is chasing the herd. In the picture a mother zebra attacks a lion, giving her defenseless foal time to escape.

Defense within the species

Many animals of the same **species** fight among themselves. Males often do this during the **courtship** season, but they very rarely kill each other. To avoid being badly hurt, such animals need some form of defense. The defenses can be either physical, which means the way in which the animals' bodies are made, or behavioral, which means the way animals react to one another.

Strong animals with sharp horns or teeth may need to be protected physically from one another. A good example is the mountain goats of North America. Two females challenge each other to find out which is the strongest. The winner will have the best pasture for herself and her young. The goats move around in circles, thrusting with their sharp horns at the other's sides and rumps. They have thicker skin in the target areas to reduce the chance of serious injury.

In the Italian Alps, two male European ibex tussle to establish which is the stronger. Their thick ridged horns absorb the powerful blows, protecting each animal from serious injury.

Fights between male kangaroos are usually very short. The loser will turn and run away before he is badly hurt by his opponent's powerful kicks.

The males of some species fight during the period of courtship, which leads up to the mating season. The purpose of this is to find the strongest male, which will then lead the herd and breed with the females. Male European mountain goats called ibex lock together their huge horns as they wrestle. However, the thickness of the horns absorbs the shock of the blows and the animals are rarely hurt. Male elephant seals try to bite each other's neck as they rear up, face to face. The thick skin around their necks helps to protect them.

Behavioral defenses are most often found among animals with the most fearsome weapons. Instead of fighting with their bodies, they use sign language. Gannets and cormorants, both fishing birds with very sharp, powerful beaks, could easily kill one another. Instead, to solve disputes over **territory**, they make signals or gestures with their beaks and wings and so avoid being hurt.

Adult male gorillas are very strong creatures but they beat their chests and roar at each other instead of fighting. They rarely hurt each other.

Animals we need to defend

Most animals have **evolved** some form of defense against their enemies during the many millions of years they have been under attack. However, **evolution** is a slow process. It takes a species a long time to become adapted to a new danger. Humans started hunting animals about half a million years ago, and in this relatively short time, animals have not evolved defenses against their new predators.

Large animals like rhinoceroses, tigers, lions and elephants have no natural predators. When they are very small they face the greatest danger, but their parents protect them. The great whales are too large to have enemies, except for killer whales that occasionally attack them. On a much smaller scale, hedgehogs are defended by their spines and toads by their poison glands. Yet all these creatures are in danger of being killed by people with bullets, **harpoons** or cars. Many animals have been nearly wiped out, or have actually become **extinct**, because they have no defense against us.

This colorful, spiny Indonesian caterpillar can successfully use its spines to defend itself against its natural enemies, such as birds. However, it has no defense against humans if they want to kill it.

Some insects, such as flies and mosquitoes, carry serious diseases. Others, such as locusts, eat our crops or infest our buildings, as fleas do. These are pests and have to be killed when they harm people.

Most insects, however, and "creepy crawlies" like spiders, worms and woodlice are harmless or even useful, but many people squash them needlessly.

At last we are beginning to understand the need to protect animals against ourselves. The great wild animals can be protected in game reserves and national parks. Smaller creatures also need help. In some countries road signs warn drivers to avoid running over toads, while in Britain some special underpasses are being built so that badgers do not have to cross busy roads. These are some of the practical ways in which we can help animals that have no defense against us.

Many animals are killed on our roads because they have no defenses against cars traveling at high speeds. In England special tunnels are sometimes built under the road for the animals to use. In the illustration two badgers have crossed a busy road safely by using a special underpass.

Glossary

Acute Sharp or sensitive.

Camouflaged Having a colored or patterned body that blends in with its surroundings.

Courtship The process during which males and females select a partner with which they will mate.

Evolution The natural process during which animals change over millions of years to suit the changing environment.

Evolved Developed to suit a particular purpose.

Extinct Never to be seen on this planet again because the last remaining member of the species has died.

Glands Body organs that release various substances, for example sweat or poison.

Harpoons Barbed spears used in whale hunting.

Mammals Warm-blooded animals, often with a hairy skin, whose young are fed with milk from the mother.

Mobbing Attacking a predator as a group.

Molt To shed feathers, fur or skin.

Nestlings Young birds that have not yet left the nest.

Nocturnal Active at night and normally resting during the daytime.

Plankton The mass of tiny plants and animals that floats in seas and lakes.

Predator An animal that kills other animals for food.

Prey An animal that is hunted by another for food.

Reptiles Cold-blooded animals, often with scaly skin, whose young are born in eggs. Snakes, lizards, tortoises, turtles and crocodiles are all reptiles.

Species A group of animals that can breed with one another, producing young that will also be able to breed together.

Territory An area in which an animal, or a pair of animals, lives and breeds.

Picture acknowledgments

The publishers would like to thank the following for allowing their photographs to be reproduced in this book: Bruce Coleman Limited 15 (C B & D W Frith), 20 (John Cancalosi); NHPA 7 (Peter Johnson), 19 (Stephen Dalton); Oxford Scientific Films 14 (G I Bernard), 27 (Kathie Atkinson); Survival Anglia 13 (Jen and Des Bartlett), 28 (Dieter and Mary Plage).

Further information

The following books will tell you more about how animals defend themselves:

Animal and Plant Mimicry by Dorothy H. Patent. Holiday, 1978.
Animal Disguises by Gwen Vevers. Merrimack Publishing Circle, 1982.
Animals in Disguise by Peter Seymour. Macmillan, 1985.
Animals That Live in Shells by Dean Morris. Raintree Publishers, 1977.
The Answer Book About Animals by Mary Elting. Putnam Publishing Group, 1984.
Amazing Facts About Animals by Gyles Brandreth. Doubleday, 1981.
Hunters and the Hunted: Surviving in the Animal World by Dorothy H. Patent. Holiday, 1981.
Poisonous Snakes by George S. Fichter. Franklin Watts, 1982.
Secrets of Animal Survival by Donald J. Crump, ed. National Geographic, 1983.
Wildlife on the Watch by Mary Adrian. Hastings, 1974.

You can discover more about how animals defend themselves by watching some of the wildlife documentaries on television, and by watching how animals behave in the countryside or even in a surburban backyard.

If you would like to help protect wild animals and the habitats they depend on, it is worthwhile joining one of the organizations listed below:

Audubon Naturalist Society of the Central Atlantic States
8940 Jones Mill Road
Chevy Chase, Maryland 20815
301–652–9188

The Conservation Foundation
1717 Massachusetts Avenue, N.W.
Washington D.C. 20036
202–797–4300

Greenpeace USA
1611 Connecticut Avenue, N.W.
Washington D.C. 20009
202–462–1177

The Humane Society of the USA
2100 L Street, N.W.
Washington D.C. 20037
202–452–1100

The International Fund for Animal Welfare
P.O. Box 193
Yarmouth Port, Massachusetts 02675
617–362–4944

National Wildlife Federation
1412 16th Street, N.W.
Washington D.C. 20036
202–797–6800

Index

Africa 6, 10, 13, 20, 21
Antelope 6, 25
Ants 16
Armadillos 13
Armor 12–13
Australia 8, 10, 15, 22, 25

Badgers 29
Bees 16, 22, 26
Beetles 6, 14
Behavioral defenses 26–7
Birds 6, 8, 14, 19, 22
Britain 29
Butterflies 19

Camouflage 8–9, 14
Canada 22
Caterpillars 8, 11, 28
Central America 7
Chemical defenses 22
Chimpanzees 15
Color, use of 8, 14, 15
Cormorants 27
Courtship fights 26–7
Crocodiles 12

Deer 8

Echidnas 10, 25
Electric shocks 16
Escape techniques 6–7
Europe 10, 21, 26, 27
Eyespots 19

Fish 4, 11, 12, 16, 22
    electric eels 16
    minnows 22
    puffer 11
    sticklebacks 11
    torpedo rays 16

Frogs 6, 8, 16
    Darwin's 20
    poison-arrow 16

Gannets 27
Giraffes 24
Gorillas 27
Group defense 22–3

Hedgehogs 10, 28
Human beings 4, 24, 28–9

Insects 6, 8, 16, 19, 29

Kangaroos 22, 24, 27
Kookaburras 22

Lapwings 21
Lizards 8, 10, 12, 13, 18
    armadillo 13
    basilisk 7
    blue-tongued skink 15
    frilled 15
    leaf-tailed gecko 8
    moloch 10
    shingleback 13
    Texas horned 10
Lobsters 7

Mobbing 22
Moths 8, 14, 16, 18
Mountain goats 26–7
Musk oxen 22–3

Opossums 20
Ostriches 21, 24

Pangolins 13
Physical defenses 26
Platypus 25

Playing dead 20
Polar bears 7
Porcupines 10
Possums (*see* opossums)
Prairie dogs 22

Rabbits 4, 6, 22
Rhinoceroses 28

Sea cucumbers 25
Seals 7, 27
Senses 4
Skunks 16
Snails 12
Snakes 8, 20, 21
South Africa 13, 19
South America 13, 16, 20
Spiders 19
Starfish 19

Tail shedding 18
Toads 15, 16
Tortoises 12

Warning calls 22
Wasps 16
Weapons
    acid 16
    hoofs 24, 25
    horns 24, 26–7
    poison 16–17, 25, 28
    spines 10–11, 28
    stings 16, 24
    teeth 24, 26
Willie wagtails 22

Zebras 24–5